AN·PELT
KENWO
WHEELED
COACH
450
SUPER DUTY
WARD LAFRANCE
ORELAND
GORTER WATER TOWER
PAT.No557799 PAT.No580084
PAT.No645470 PAT.No708135
PAT.No718965.
GELES
CROWN
Firecoach
Peterbilt
REIGHTLINER
AMERICAN
LaFRANCE
White
ERICAN LAFRANCE
LTi
SHKOSH
STUTZ

Horses to Horsepower

A Pictorial History of the Apparatus of the Los Angeles City Fire Department

Compiled by
Dave Cox and Walt Jaeger
for the
Los Angeles Fire Department Historical Society

Table of Contents

Dedication 3
Los Angles Fire Department Historical Society 4
Introduction 5
Fire Horses 6
Chapter I Horse Drawn and Early Motorized Vehicles, 1886 to 1919 ... 7
Chapter 2 Vehicles of the 1920s, 1930s, & 1940s 13
Chapter 3 Mid-Twentieth Century Apparatus, 1950s 39
Chapter 4 Apparatus of the 1960s 63
Chapter 5 1970s Apparatus 107
Chapter 6 1980s, the Last of the Open Cabs 135
Chapter 7 End of the 20th Century, 1990s 149
Chapter 8 Into the 21st Century, 2000 on 169
Chapter 9 Apparatus Models 195
Acknowledgements 206
Biblography 206
Index 208

M.T. Publishing Company, Inc.™

P.O. Box 6802, Evansville, Indiana 47719-6802
www.mtpublishing.com

Printed in the United States of America.

Library of Congress Control Number:
2012941596

ISBN: 978-1-934729-87-8
Graphic Design by:
Alena Kiefer,
M.T. Publishing Company, Inc.

Front cover photos: CSFA 1902 American Metropolitan steamer similar to those used by the LAFD, and two 1996 Seagrave triples. Both photos by Walt Jaeger. Back cover photo: LAFD fire scene photographed by Mike Meadows.

Dedication

This album of historic apparatus of the Los Angeles Fire Department would not be possible if it were not for the thought, effort, and ability of the many photographers and historians who have recorded these images and information over the years. We are grateful for the courage, patience, and persistence of photographers like Chuck Madderom, Mike Meadows, Keith Cullom, Larry Arnold and so many others, who obtain the perfect pictures at the right moments, and also to those historians who searched records to verify every aspect of an apparatus' existence; respected and knowledgeable people like John A. Ackerman, Walt Pittman, Dale Magee, Robert Allen, and Paul Ditzel.

We dedicate this album to all, through whose efforts we were able to assemble this historic collection for the LAFD Historical Society.

This 1923 Seagrave 1000 gpm triple was restored to its original appearance by Historical Society members Mark Howell, shown driving, Tim Griffin, and Leo Dempsey. It was assigned to Fire Station 49 and is now on display at the LAFD Historical Society Museum in San Pedro.

A pair of 1923 Seagrave triples out of service in 1960.

Originally a chemical hose wagon with a 40 gallon soda acid tank, this 1923 Stutz was modernized with a 165 gallon water tank and a 75 gpm booster pump. The Stutz was one of the first apparatus purchased to come with pneumatic tires. It served with a 1923 Seagrave Triple at Fire Station 49.

A 1923 Stutz Salvage Wagon assigned to Fire Station 28.

This 1923 Seagrave 1000 gpm double combination was restored and is now on display at the LAFD Historical Society Museum in San Pedro, which was formerly Fire Station 36.

This retired Stutz straight frame tillered city service truck was one of five purchased in 1923 and 1924.

Truck 2 was a 1923 Seagrave 85 foot aerial that was modernized with a windshield and pneumatic tires. The two large boxes on the side contain Gibbs Re-breathers.

This 1923 Moreland Foam Carrier, one of four built by the department's shops, carried 112 fifty pound cans of dry foam power and other equipment for fighting oil fires. It is shown here, no longer in service, in the 1960s.

The 1923 Seagrave truck had a lifenet in a box mounted on the rear but no tiller windshield.

Fire Boat 2, the Ralph J. Scott, cruises the harbor in the 1960s. Built in 1925, Boat 2 was, for many years, the most powerful apparatus in the department.

Boat 2 putting on a display in the harbor main channel.

This retired 1930 Moreland Personnel Carrier was photographed at the LAFD Shops in 1960. The truck is currently under restoration by the L.A. County Fire Museum.

A 1929 Seagrave Chemical Hose Wagon that was converted to a dry-powder foam wagon.

This hose carrier was originally a 1925 Mack Salvage Wagon. It was converted in 1930 to a hose wagon. A Hall-Scott engine was installed along with front-end parts from various other apparatus.

The LAFD used this 1930 Moreland Brush Tanker, with a body built by Standard Auto Body Works, with a 150 gpm pump and a 650 gallon tank into the 1960s. Two of these rigs were purchased by the department in 1930.

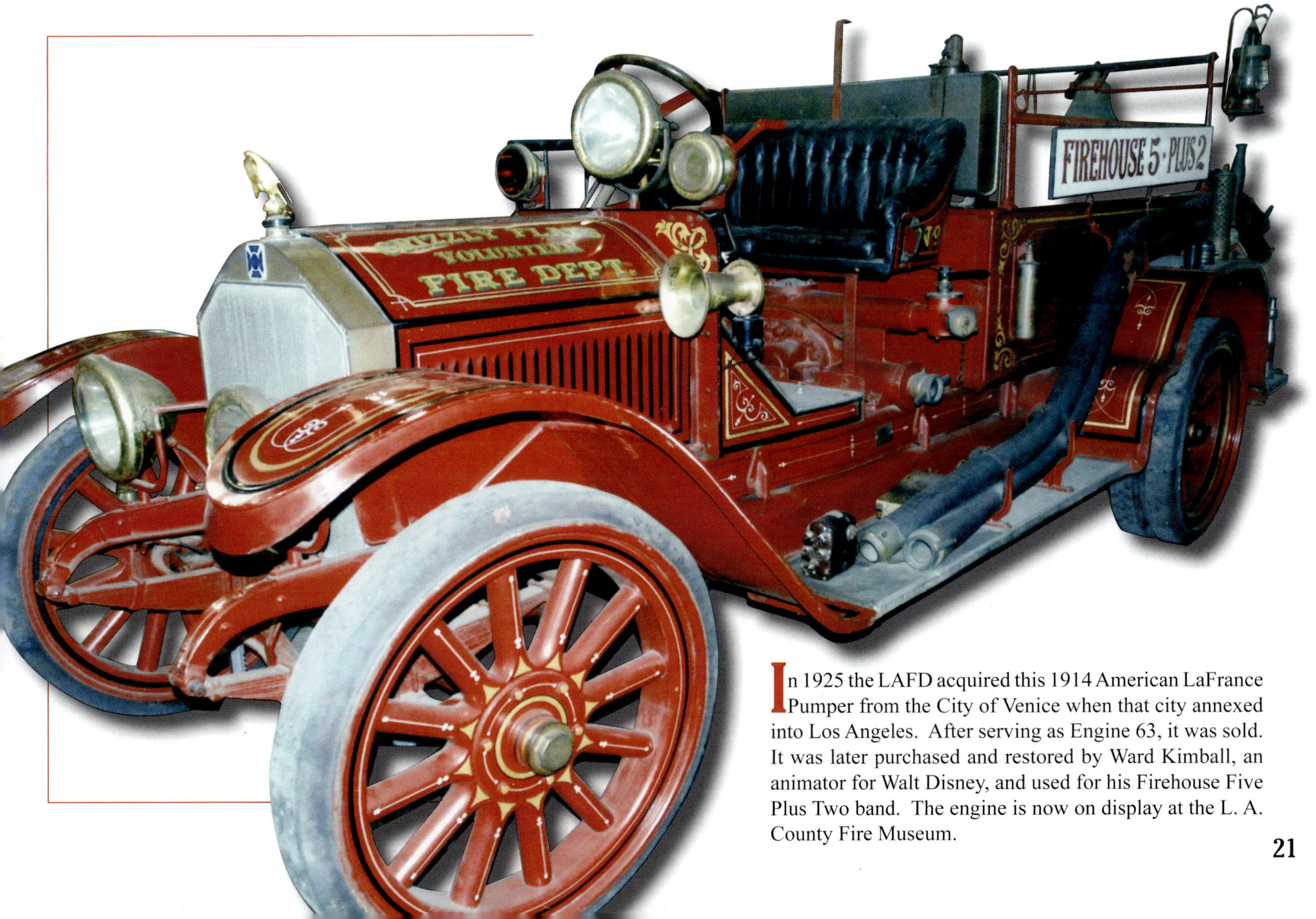

In 1925 the LAFD acquired this 1914 American LaFrance Pumper from the City of Venice when that city annexed into Los Angeles. After serving as Engine 63, it was sold. It was later purchased and restored by Ward Kimball, an animator for Walt Disney, and used for his Firehouse Five Plus Two band. The engine is now on display at the L. A. County Fire Museum.

This 1931 Seagrave Hose Wagon was restored as a parade vehicle with seats in the hosebed for the LAFD band and was nicknamed the "Bandwagon."

A 1932 Seagrave Hose Wagon in use in the 1960s. The original 60 gallon water tank was replaced by a 170 gallon water tank in the 1940s.

This 1937 American LaFrance Duplex 2000 gpm rated pumper was still in front line service in 1960. These pumpers had two 1250 gpm motor/pump units, one in front and the other in back of the cab.

A street side view of a 1937 American LaFrance Duplex Pumper. Four of these pumpers were delivered to the department: two 2000 gpm units in 1937 and two 3000 gpm units in 1938.

One of two closed-cab 1938 American LaFrance 400 series city service trucks operated by the LAFD. These were the last city service trucks purchased by the department due to poor ratings by the insurance underwriters.

This 1937 American LaFrance Triple with a fully enclosed cab is being restored at the LAFDHS Museum in Hollywood. Five of these triples were purchased with 1250 to 1500 gpm pumps and 80 gallon water tanks. They were nicknamed "lulubelles" since they were big and clumsy like a cartoon character of the time.

The rear hose bed on the manifold wagon had 1200 feet of 3 1/2 inch hose and the transverse hose beds contained 1500 feet of 2 1/2 inch hose and 200 feet of 1 1/2 inch hose. The wagon also had an 80 gallon water tank.

A 1938 American LaFrance Manifold Wagon that ran with a duplex pump.

This Seagrave Manifold Wagon ran with an American LaFrance Duplex Pumper as shown here at Fire Station 9 in 1954.

In 1938 the LAFD bought two of these Seagrave Manifold Wagons. They carried 2700 feet of hose, had six water inlets and 18 outlets, and were supplied water by a duplex pumper.

Here, the 1938 American LaFrance Manifold Wagon and a duplex pumper are responding out of Hollywood's Fire Station 27 in 1960.

The Los Angeles Municipal Airport was protected by this 1941 Mack Crash Truck. The truck has a 200 gpm pump and a 350 gallon tank. This truck has been restored and is now privately owned.

This rear view of the Mack Crash Truck shows the hose bed and designation as Crash 80.

This 1943 Kenworth tractor was coupled to a 1938 Seagrave aerial in 1960.

In 1941 the LAFD bought one of these open cab United-Kenworth 1250 gpm triples with a 200 gallon tank. It was in service for nearly 20 years.

A 1945 Pirsch 1250 gpm triple with a 200 gallon tank. The LAFD bought 18 of these Pirsch Triples in 1945. They were the first engines to have the suction connection at the front of the vehicle.

This 1945 open cab American LaFrance 500 series 100 foot aerial was the first metal aerial purchased from American LaFrance.

A 1929 Seagrave Triple that was refitted with a windshield and a 165 gallon water tank.

A 1945 Seagrave 85 foot aerial in reserve service at the Valley Shops behind Fire Station 88.

This 1945 Kenworth 1250 gpm triple had a 200 gallon tank and a one inch booster reel mounted at the rear step.

This 1923 Seagrave Triple was modernized with a windshield in the 1940s. The 60 gallon chemical tank was converted to a water tank.

The Van Nuys Airport was served by this 1944 International dry powder crash truck. It was nicknamed the "duster."

Mountain Patrol used this 1947 International semi to haul a skip loader.

Background photo:
Apparatus from Fire Station 27 at a garage fire in Hollywood.

This fuel tender was built on a 1948 GMC Salvage Wagon chassis.

KENWORTH
984
LOS ANGELES
CITY
FIRE DEPARTMENT
CLASS 1

F40
E107997

Opposite page:

In 1948 the LAFD bought this Kenworth Heavy Utility. It featured a Holmes twin boom with a ten ton capacity crane, compressor, and cutting torch supplies.

Bottom Photo

1949 Seagrave 1250 gpm triple with a 200 gallon tank; one of 15 purchased in 1948 and 1949. This engine had been converted to a foam carrier at Fire Station 40 on Terminal Island in the Port of Los Angeles.

In 1950 the LAFD bought two of these Pirsch 85 foot aerial trucks.

A 1951 Kenworth 6x4 tank wagon with two 300 gpm pumps and a 2500 gallon tank. This was the largest tanker at that time and was designed for brush fire work. It later saw service at the Van Nuys Airport as a nurse tender before being on reserve at Fire Station 88 in the Valley.

One of two 1951 Mack Tank Wagons with 300 gpm pumps and 700 gallon tanks.

A 1953 Mack 1250 gpm triple with a 200 gallon tank. Four of these triples were delivered.

The only Seagrave Combination purchased was this 1951 model. It featured a 1000 gpm pump, wagon battery, hose, and ladders but no water tank. It was designed for use at Terminal Island.

In 1951 this Mack Heavy Utility was put in service at Fire Station 27 and later moved to Fire Station 88.

A rear view of the Mack HU showing the Holmes twin boom crane.

This view shows the various fittings carried by the 1952 Mack Boat Tenders.

Two 1952 Mack/Yankee Boat Tenders were assigned to the Port of Los Angeles fire stations. These trucks had large boat size monitors and carried 1350 feet of 3 1/2 inch hose.

This 1953 GMC Tank Wagon carries 400 gallons of water.

A rear view of the GMC Tank showing the outlets.

A 1953 Seagrave 85 foot aerial powered with a V-12 motor. This truck was a demonstration model purchased by the city, used, in 1957.

A light utility built on a 1948 GMC body by the Shops in 1953. The light wagon had 8 lights mounted on a telescoping stand, portable floodlights, cable reels, and a 30,000 kw generator.

The first Crown Triple purchased was this 1953 model with a 1250 gpm pump and a 350 gallon tank. (F-1005)

This 1953 GMC was converted from a booster tank to a compressor/demolition truck.

A 1953 Kenworth/Yankee Tank Wagon with a 600 gallon tank and two 150 gpm pumps.

This classic 1954 Cadillac ambulance is on display at the LAFD Historical Society Museum in Hollywood. Two of these ambulances were in service, at Fire Stations 3 and 27.

Here is a 1954 Crown 1250 gpm triple with a 350 gallon tank and shop-added transverse hose beds. (F-1013)

A 1955 Seagrave Anniversary Series 1250 gpm triple. The hood side panels were removed to help cool the engine.

The crew of this 1955 Seagrave Triple attempt to protect exposures at a structure fire with their wagon battery.

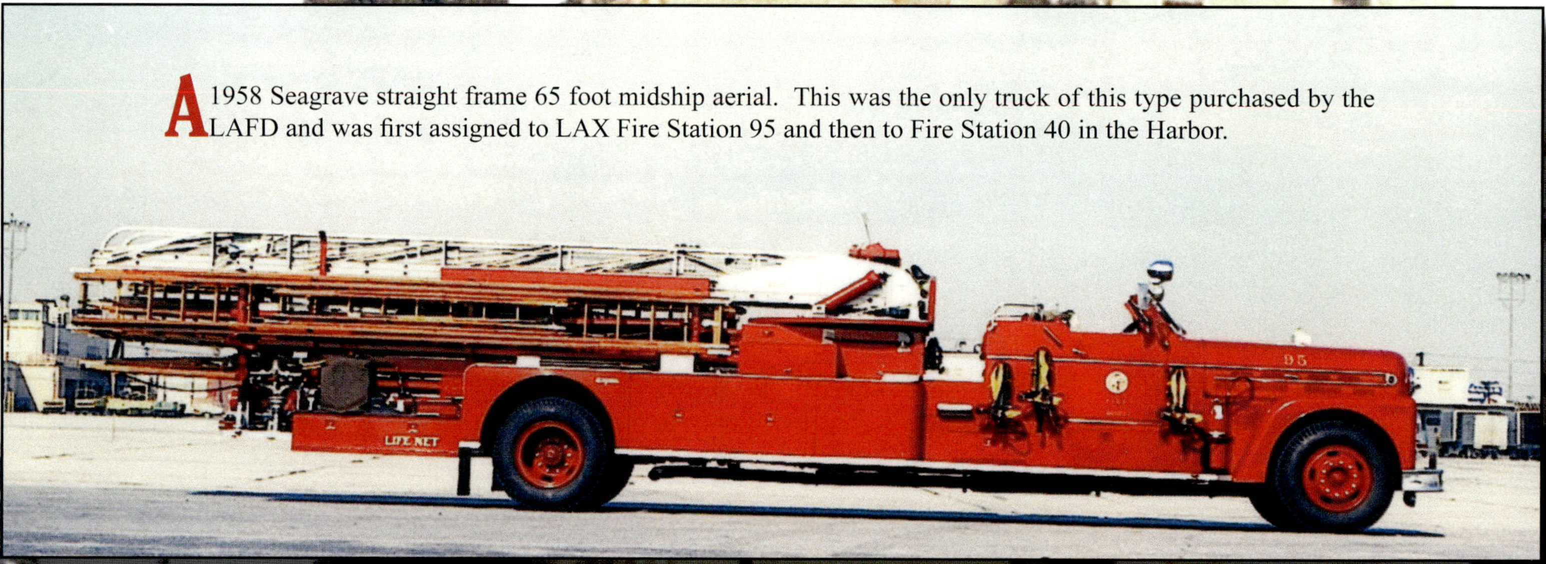

A 1958 Seagrave straight frame 65 foot midship aerial. This was the only truck of this type purchased by the LAFD and was first assigned to LAX Fire Station 95 and then to Fire Station 40 in the Harbor.

A 1938 Seagrave aerial from Fire Station 3 attends to this commercial fire in the downtown area.

A front view of the Seagrave midship aerial.

This is Truck 66, a 1955 Seagrave 85 foot aerial. Fire Station 66 is seen in the background.

Background photo:
The 1957 FWD/Yankee Crash truck in action at a drill site fire.

A 1957 FWD/Yankee 6x6 750 gpm crash truck with a 1500 gallon water tank and 220 gallons of foam. This unit was first assigned to LAX and then later to Fire Station 90 at Van Nuys Airport.

Crash Tanker 95 is a 1958 FWD/Yankee 6x6 with a 750 gpm pump and a 2500 gallon tank. It was assigned to Los Angeles International Airport (LAX).

The LAFD purchased five 1955 Crown High-pressure Hose Wagons with two 150 gpm high-pressure pumps and a 400 gallon tank. (F-1036) and (F-1037 to F-1040) The enclosed transverse hose beds contained 1 1/2 and 2 1/2 inch hose.

A 1957 Ford Communications van shown here in front of Fire Station 69 in Pacific Palisades.

A 1957 Crown as it was later converted to a foam carrier and assigned to Fire Station 100 in Van Nuys. (F-1076)

A 1957 Crown High-pressure Wagon with a two 150 gpm pumps and a 400 gallon water tank. (F-1075)

A lot of hose or "street spaghetti" is seen in this business district fire.

This 1959 Crown Triple with a 1250 gpm pump and a 400 gallon tank was one of five purchased that year. (F-1161 shown), (F-1157, F-1158, F-1160 and F-1167)

In 1958 the department bought five of these Mack 1250 gpm triples with 400 gallon tanks.

A 1956 Seagrave Anniversary Series 1250 gpm triple with a 400 gallon tank. It was powered by a V-12 engine.

A classic 1958 Buick Battalion Chief's car.

This 1957 Ford, converted to a helicopter tender, operated at Van Nuys Airport.

Dozer with a hose reel overhauling at a fire scene.

A 1957 GMC Rescue Ambulance in front of Fire Station 27.

One of three 1959 FWD 4x2 high-pressure hose wagons with two 150 gpm pumps and a 400 gallon tank.

Below:

One of two 1958 Pirsch 85 foot aerial trucks purchased that year.

Chapter 4

Apparatus of the 1960s

To help protect the expanded city of over 470 square miles, the LAFD purchased its first helicopter in 1962, after the Bel Air brush fire. Helicopters were able to serve in the capacity of observation, water dropping, rescue, transportation of injured and department personnel. Apparatus design changed after the Watts riots in 1965 when fire personnel were being injured by thrown rocks and bottles in their open cab trucks. The department shops built solid roofs and "riot cages" of steel mesh around the cabs. Some new apparatus were ordered with solid roofs on the cabs. These roofs would also be beneficial in protecting the new electronics in the cabs.

The Manifold-Duplex System was unique to Los Angeles and continued in service until 1964 when they were replaced by a 2000 gpm Crown Pumper and a Crown Manifold Wagon assigned to Fire Station 17.

Fireboat 4, the Bethel F. Gifford, was added to the fleet in 1962. The 76 foot boat was built by Albina in Portland, Oregon. In 1967 three smaller fireboats were purchased. These 34 foot Drake Craft boats served as Boat 1, 3, and 5.

Crown Firecoach advertisement in the 1960s. Los Angeles was a major customer of Crown fire apparatus.

A 1960 International Salvage Wagon at Fire Station 5 in Westchester.

A closed-cab 1960 Crown tractor transport hauling an Allis-Chalmers bulldozer. The Crown tractor transport was destroyed in a brush fire in 1967. (F-1223)

This 1948 GMC Salvage Wagon was assigned to Fire Station 61 in the 1960s.

As a modernized 1929 American LaFrance 1000 gpm triple, this engine was in reserve service in 1960. The windshield and a Hall-Scott engine were added in the 1940s.

A 1960 Ford Mountain Patrol truck.

This 1960 Ford Commissary, or food service, truck was built in the LAFD Shops and used to feed firefighters at the scene of extended incidents.

A 1923 Seagrave 85 foot wooden aerial in service in 1960.

Seagrave

This 1960 Seagrave 100 foot aerial truck was the first cab forward design purchased from Seagrave.

A 1960 Seagrave aerial truck with a ladder pipe in operation at a structure fire.

The restored 1905 Gorter Water Tower being housed at Fire Station 7 on South Main Street in the 1960s.

A 1960 Crown 1250 gpm triple with a 400 gallon tank. (F-1194) Crown delivered 26 of these triples in 1960. Note the top mounted transverse hose beds.

STATION NO 90

LAKE ARROWHEAD
FIRE DISTRICT

Only one 1960 Pirsch 100 foot aerial was purchased.

Opposite page:

In the 1960s, this former LAFD 1929 American LaFrance 1000 gpm pumper was still in use in the community of Lake Arrowhead in the San Bernardino Mountains.

This arson investigation unit was built on a 1960 Ford chassis formerly used as a commissary truck.

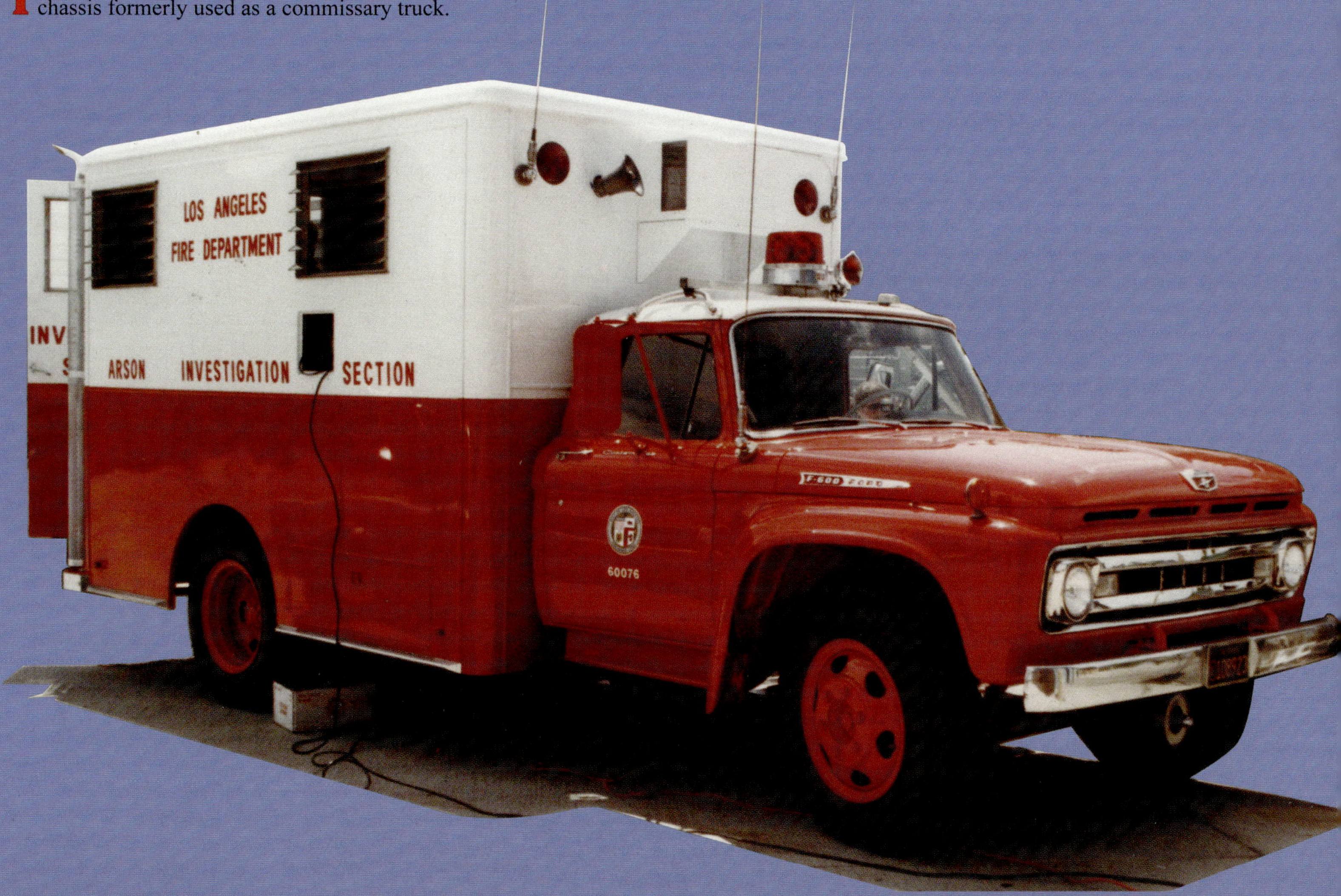

Fire Boat 4 shown as delivered in 1962. The boat is 76 1/2 feet long with six engines and a pumping capacity of 9000 gpm.

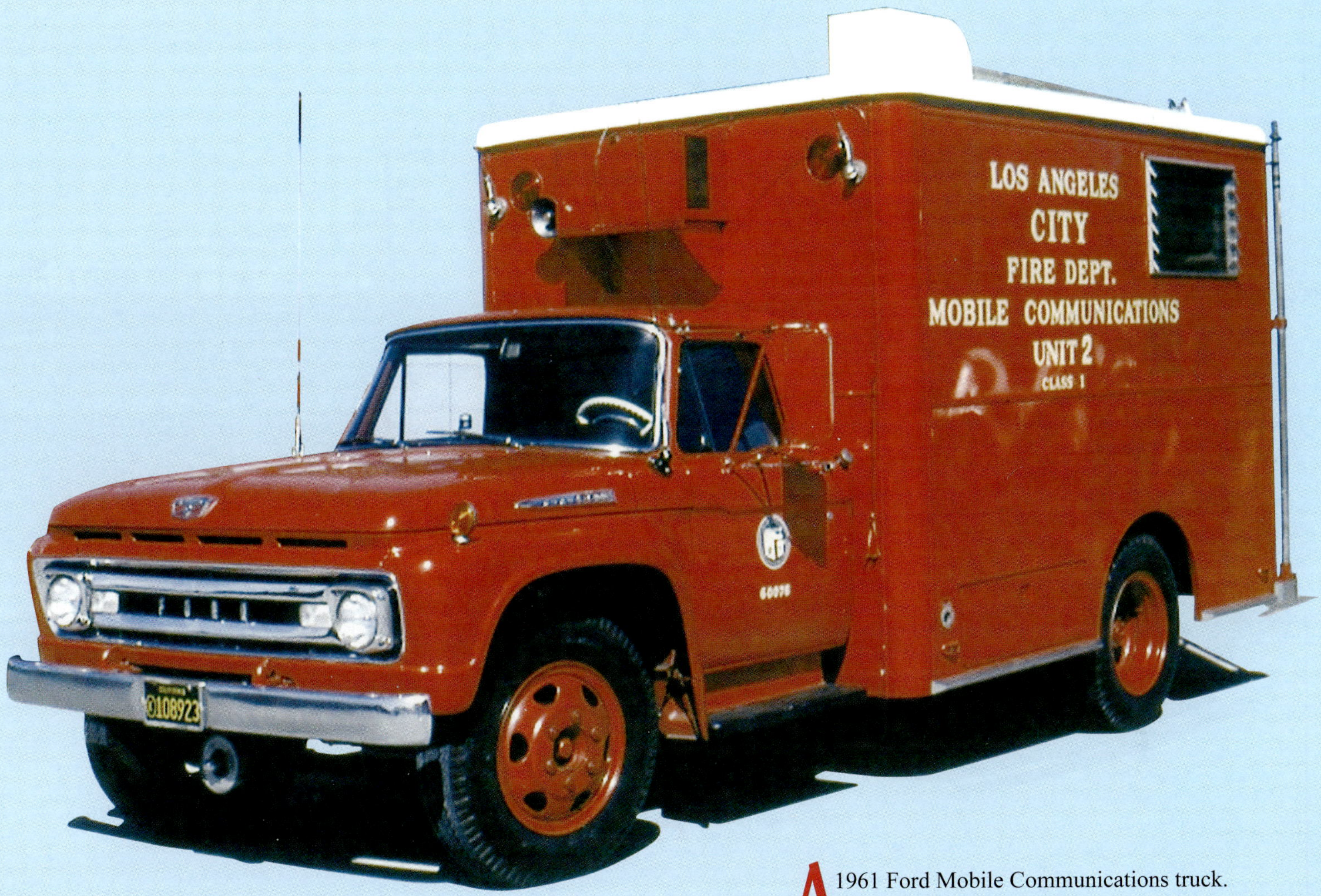

A 1961 Ford Mobile Communications truck.

A 1962 Seagrave 100 foot aerial.

A 1961 Crown Triple with a protective cab cover. (F-1256)

Engine 8, a 1961 Crown, at a house fire in the Porter Ranch area of the San Fernando Valley. (F-1259)

This 1962 Seagrave 100 foot aerial ran out of Fire Station 47 in El Sereno.

This 1962 Chevrolet food service utility was donated to the LAFD by the grateful residents after the Bel Air fire and is still in use by the LAFD Historical Society.

A 1964 Bell 206 Jet Ranger helicopter used for observation.

A 1962 GMC/Yankee Tank Wagon with a 150 gpm pump and a 400 gallon tank.

The department picked up this three passenger Bell 47G-3B helicopter from the factory in Texas in 1962. Unfortunately, it crashed on June 23, 1974.

Shown here is a 1962 tilt cab GMC 300 gpm tank wagon with a 600 gallon tank and an overhead ladder assigned to a mountain area station. The tank wagon was equipped with a 35 foot extension ladder since the mountain areas had no truck companies. The LAFD purchased eight GMC tank wagons in 1962.

A 1962 Crown 1500 gpm triple with a 400 gallon tank and a Stang wagon battery. (F-1312)

Since this 1962 Crown 1500 gpm triple served the Los Angeles Airport, the FAA required that it be painted yellow. (F-1311)

This 1962 Crown 1500 gpm triple, also painted yellow, was not assigned to an airport, but served the San Fernando Valley at Fire Station 81.

A 1960s Seagrave truck working a fully involved furniture store fire.

This 1962 GMC stake-bed truck pulled a search and rescue trailer.

A 1964 Dodge Patrol truck was used in the brush areas in the hills of the city.

Fire Station 89 housed this fuel truck built on a 1963 GMC truck chassis.

A 1960s Seagrave triple's wagon battery in use at a working fire.

A 1963 Chevrolet/Yankee Salvage Wagon.

The "Big Bertha" wagon battery of Fire Station 17's Crown Manifold in action at a restaurant supply fire.

Above:

This 1963 Crown Triple had an elevated ladder rack with a 35 foot extension ladder. (F-1335) Since this triple served the Terminal Island area, where there were no aerial trucks, the extra ladders were needed.

A close-up of the Intelegiant wagon battery on top of the 1963 Crown Manifold.

Below:

Wagon 17 was a 1963 Crown Manifold with a 2000 gpm pump, 400 gallon tank, and a hydraulically operated wagon battery, nicknamed "Big Bertha." (F-1338) This monster rig is currently on display at the LAFDHS Museum in Hollywood.

One of the crash rigs that protected LAX was this 1964 FWD/Yankee 4X4 Rapid Intervention Vehicle. It carried 1000 pounds Purple K dry chemical, 500 gallons of foam, and 375 pounds of CO_2. It also had an extendable light tower and a 5000 watt generator.

The LAFD received its second helicopter, a Bell 47 J-2 four seat Rotocraft, in 1964.

A 1964 Walter/Yankee 4x4 crash tanker with a 750 gpm pump, 2500 gallons of water, and 1000 pounds Purple K dry chemical.

Background photo:
LAFD apparatus pour water on arson-set fires during the Watts riots of 1965.

A 1963 Crown 2000 gpm triple with a 400 gallon tank. This engine operated with and supplied the 1963 Crown Manifold. (F-1339) It was the pump for the two piece engine company at Fire Station 17.

A 1965 GMC/Yankee 150 gpm patrol tank with a 400 gallon water tank.

This 1966 Seagrave 100 foot aerial ladder truck was assigned to Fire Station 9.

Ten of these 1964 Seagrave 1500 gpm triples with 400 gallon water tanks were ordered.

Above:

A 1960 Crown with shop-built "riot cab" resulting from the Watts riots in 1965. These screened enclosures were built at the LAFD shops at a cost of $175 per vehicle. They provided top and side protection from thrown rocks, bricks, and bottles. (F-1208)

Right:

This 1967 Ford/Yankee two man squad was evolved from a salvage truck.

Two of these 1967 GMC/Yankee Salvage Wagons were delivered.

A covered-cab Seagrave ladder at a hot structure fire.

A 1967 Ford fire patrol. These trucks were equipped with a pump, a small water tank, and hose.

This rescue ambulance was a 1967 Ford.

This 1949 Mack Manifold wagon had ten 2 1/2 inch outlets and carried 3 1/2 inch hose in back and 2 1/2 inch hose in the enclosed transverse beds.

One of four Dodge/Yankee salvage trucks purchased in 1964 and 1965.

Fireboat 1, the Archibald Eley, purchased in 1919, originally had no pilot house. It is shown here with a custom built pilot house just before the end of its service in 1968.

An elevated view of a two-piece engine company, wagon and pump, consisting of Crown Triples.

1967 Drake 34 foot fire boat with a 750 gpm pump at Fire Station 111. Two of these Drakes were purchased in 1967 and a third one in 1968.

Fireboats 4, 5, and 2 at a wharf fire in 1967. Boats 4 and 2 are advancing on the fire while Boat 5 is acting as a command post.

A 1967 Crown 1500 gpm triple with a 400 gallon tank. (F-1460)

This 1967 GMC bus was used to transport additional firefighters to the scene of major incidents.

This Caterpillar diesel tractor was used for clearing brush in the hillsides and mountains of Los Angeles.

A 1968 Walter/Yankee Crash Rescue with a 3000 gallon tank. This rig had two 750 gpm pumps and two gasoline motors. Three of these units were in service at LAX, Crash 1-80, 2-80, and 3-80.

A 1968 Kenworth diesel transport with a 1966 Caterpiller bulldozer.

All of LAX's fire apparatus gathered for this group photo.

A 1968 Crown 85 foot Snorkel assigned to Fire Station 3. (F-1500) The use of the 85 foot Snorkel was abandoned in 1970 after an accident.

A rear view of the 85 foot Snorkel 3.

A 1969 International diesel transport with a 1970 Caterpiller D7 dozer.

This large fuel tank truck was on a 1969 Diamond "T" truck chassis.

One of two 1969 Crown 2000 gpm pumpers purchased without water tanks. (F-1572)

The same 1969 Crown engine, converted into a triple by the shops in 1984, with a 500 gallon water tank and a diesel engine.

One of two 1969 Crown/Maxim 75 foot aerials delivered with Hall-Scott gasoline engines. The shops added an additional fly to the ladder to lengthen them to 100 feet and re-powered them with diesels. (F-1585)

This 1969 International was one of the early types of rescue ambulances.

The Los Angeles Police Department used these 1969 Ford ambulances before the fire department ran the ambulance service in all parts of the city. They were called "Brown Bombers" and responded out of the police stations throughout the city. They were repainted red for the LAFD service.

The LAFD bought two of these 1500 gpm Crown Snorkels with 300 gallon tanks in 1969. (F-1570) (F-1571)

Fire Boat 4, a 1961 Albina, with new red and white paint job.

Chapter 5

1970s Apparatus

In 1970 the first paramedic ambulance went into service at Fire Station 53 in San Pedro. Today the goal is to have at least one rescue ambulance in each station. After the 1971 earthquake the LAFD learned many lessons about how search and rescue operations could be improved. The department joined the National Institute for Urban Search and Rescue along with other organizations. In 1978 the LAFD put the first mobile lab in service. The mobile lab was a converted van with chemical protective suits and limited detection equipment. These were the beginnings of the hazardous materials, or haz-mat, vehicles.

1970 Seagrave Triple as delivered without a cab cover.

A 1970 Seagrave Triple with cab cover, now privately owned.

This 1948 Mack Salvage Wagon was converted to a helicopter tender in 1970.

One of the 1948 Mack Salvage Wagon conversions, this time into an air utility unit in 1973. The light on the rear of the truck telescopes above the truck to show its location at fire scenes.

The 1948 Mack Air Utility showing the bottle storage and compressor compartments.

One of two 1970 Seagrave 100 foot aerial ladder trucks purchased with a Cummins diesel engine. These were the first diesel powered aerials purchased.

Fire Station 3's Crown Triple and GMC Rescue Ambulance ready to roll.

One of the Department's tractor transports was this 1970 Mack with a 1970 Caterpiller D7 Dozer, designated Dozer 1.

A 1970 Crown 85 foot diesel powered Snorkel assigned to Fire Station 27. (F-1622) This was the second 85 foot Snorkel purchased and was taken out of service later that year.

A 1970 Van Pelt 1000 gpm triple with a 400 gallon tank. The LAFD bought seven of these engines over a two year period.

Fire Boat 4 in action at a wharf fire.

There are hoses everywhere; under and over the engine at this fire scene.

Truck 37's 1970 Seagrave ladder at a residential roof fire.

Three of the Department's 100 foot Thibault aerial ladders were open cab models like this one from Fire Station 61.

This 1971 Crown 1500 gpm 50 foot Snorkel (F-1648) was one of two delivered to the LAFD.

A rear view of the Crown Snorkel. The apparatus served the Hollywood area of Los Angeles at Fire Station 27.

The LAFD owned two Crown 50 foot Squirts.

Here is the Crown (F-1671) with a 50 foot Telesquirt, a two section ladder, and cab roof.

Crown built this 1972 1500 gpm triple combination pumper (F-1671). It was powered by a 350 hp Cummins diesel engine.

Engine 16 (F-1649) was built from a 50 foot Crown Snorkel with an added cab roof.

Engine 213, shown here, was built from a 1971 Crown Snorkel. A new body was built and a 500 gallon tank was added. Both SN 60255 (F-1648) shown here and SN 60256 (F-1649) were rebuilt into triples.

In 1972 the LAFD bought six Thibault 100 foot aerial ladder trucks. All of the Thibault trucks were taken out of service in 1976 so their aerial ladder assemblies could be replaced with Seagrave aerials.

A 1972 Crown two section 50 foot Telesquirt with ladder, SN 60265 (F-1669). It had a Cummins 6 cyl diesel engine and a 1500 gpm Waterous two stage centrifugal pump.

This view shows the mounting for the Telesquirt in the rear of the engine. The old designation as Wagon 9 became Engine 209.

This air utility truck was converted from a 1958 Ford Salvage Wagon in 1972.

One of the first rescue ambulances to serve the Harbor area was this 1970 Ford 3/4 ton van.

Fire Station 17 responded to medical emergencies with this 1973 Chevrolet ambulance.

A 1973 Bell 205A-1 helicopter shown at the LAFD base adjacent to the Van Nuys Airport.

A Task Force portrait in front of Fire Station 29 consisting of a 1971 Crown Squirt (F-1647), a 1973 Crown Triple (F-1682), and a 1970 Seagrave 100 foot aerial.

Engine 18 ran with this 1972 covered cab Crown (F-1670) 1500 gpm triple with a 400 gallon booster tank.

The LAFD received 14 of these 1972 Ford C-8000 American LaFrance 1000 gpm triples with 400 gallon water tanks. American LaFrance built the body and fire fighting equipment on a Ford commercial chassis. They were powered by Caterpillar diesel engines.

Wagon (W) 39 was this 1974 Crown 1500 gpm triple (F-1707) with a 400 gallon tank. The wagons were part of a two- piece engine company designated as wagon and pump. Los Angeles took delivery of ten of these Crown rigs in 1973 and 1974.

Two of these 1973 Crown engines were delivered. Each triple had a 1500 gpm Waterous pump and a 400 gallon tank. Pump (P)10, pictured, was Crown number F-1682.

The Seagrave Co. delivered nine of these 100 foot aerial ladder trucks in 1975. Task Force 98 received the first one.

The 1974 Mack Heavy Utility followed by the older 1968 Crown Heavy Utility.

This 1974 AMC Matador Battalion Chief's buggy positioned in front of old Fire Station 17.

This 1959 Crown Triple (F-1167) was converted to a crash tender and served at Fire Station 90 and Van Nuys Airport.

Engine 99 (F-1700), shown here, was one of the 1973 Crowns with cab covers.

The back end of the 1959 Crown crash tender is shown here in 1974.

A Chevrolet ambulance responding out of Fire Station 34.

This 1977 Pontiac Chief's sedan is now on display at the LAFD Historical Society Museum in Hollywood.

This 1962 GMC 4x2 was a tank wagon before being converted to a helicopter tender in the 1970s.

This 1975 Seagrave ladder truck received a roof over the cab to protect the electronics and a new light bar.

A 1975 Plymouth Battalion Chief's sedan .

The LAFD received 22 of these new Ward-LaFrance Triples in 1976. They were equipped with a 1500 gpm pump and a 500 gallon water tank. At 424 hp, these motors were the most powerful ever installed in an LAFD engine to that time. They were assigned in pairs, wagon and pump, to eleven fire stations.

This 1957 Crown High-pressure Hose Wagon was converted into a foam carrier in 1977. (F-1089)

Rear view of converted foam carrier. The LW stands for Light Water but that was later changed to FC, Foam Carrier.

This photo shows one of the 1976 Ward-LaFrance engines with a covered cab and new warning lights.

This Walter/Yankee 4x4 Rapid Intervention Vehicle was stationed at Fire Station 80 at LAX. It was equipped with a 500 gpm pump and a 500 gallon water tank, 500 pounds of Purple K and 500 gallons of foam.

Ten of these 1977 Seagrave 100 foot aerial ladder trucks were put in service throughout the city. These were the last of the open cab Seagrave trucks purchased by the LAFD.

This military surplus 1964 Bell 204 helicopter was purchased in 1978 for $500.

Above:
Helicopter Tender 2, a 1979 Mack MB model, was capable of delivering 2000 gallons of jet fuel to the fleet of LAFD helicopters at Van Nuys Airport.

Below:
Fire Patrol 99 was one of six 1978 Ford patrol trucks equipped with 125 gpm pumps and 200 gallon water tanks. The patrol trucks were assigned to stations in the fire-prone hillside areas.

A firefighter is dwarfed by the size of this giant apparatus. These crash trucks are 45 feet long, are powered by two 492 hp diesel engines, and carry 4000 gallons of water, 500 pounds of Purple K, and 515 gallons of foam.

Foam 180, a 1979 Oshkosh Crash truck, spraying foam at LAX.

Two of these huge 1979 Oshkosh M-12 8x8 crash trucks with two 1250 gpm pumps were stationed at Fire Station 80 at LAX. They were designated as Foam 1-80 and Foam 2-80.

Chapter 6

1980s, the Last of the Open Cabs

By the 1980s all of the LAFD apparatus had closed cabs, and the days of riding on the tailboard were over. In 1982 three Strategic Chemical Attack Teams (SCAT) were implemented and assigned to stations 4, 27, and 39. By 1985 the SCAT name was changed to Hazardous Materials Squads. The vehicles were upgraded to 1972 American LaFrance triples, converted for that purpose. Later two large motor homes, known as Haz-Mat Response Units, were put into service. In 1987 the Disaster Preparedness Division was formed, and research began on a specialized program which led to the formation of the Urban Search and Rescue (US&R)vehicles.

This Chevrolet ambulance, with an extended roof, was assigned to Fire Station 27 in Hollywood.

This 1962 GMC high expansion foam carrier was converted from a salvage unit in 1980.

A lineup of Fire Station 80's crash units at LAX. Shown left to right; a 1980 Oshkosh/Yankee, a 1968 Walter/Yankee, and two 1979 Oshkosh/Yankee Crash Wagons.

A firefighter died at this structure fire when he fell through the roof. The building, originally part of a naval reserve center, was rebuilt and is now known as the Frank Hotchkin Memorial Training Center, named after the fallen firefighter.

Above:
The LAFD purchased 11 of these diesel-powered 1980 Seagrave 1500 gpm triples with 500 gallon water tanks.

Right:
The LAFD also had some special apparatus on its roster. This 1981 International dump truck was kept at Fire Station 88 in the San Fernando Valley and was used for a variety of assignments throughout the city.

Below:
One of the 1980 Seagrave Triples was painted yellow and was assigned to Fire Station 51 at Los Angeles International Airport.

This twin motor Chinook helicopter was used to lift the Suspended Maneuvering System. The SMS was developed by McDonald Douglas Corporation and was designed to deploy fire fighting or medical personnel to remote areas or hazardous sites.

The SMS, when suspended from a helicopter by a 1000 foot cable, could maneuver 160 feet in any direction independent from the helicopter.

This experimental Suspended Maneuvering System unit, or SMS, contained a 210 horse power propulsion engine and four thrust nozzles to move it into position. The system was never used by the department at a fire and is now in storage.

These Crown Triples were no longer in service and waiting for some future use.

Ten of these 1982 Seagrave 1500 gpm triples were purchased. This one was in service at Fire Station 68 in the mid-city area.

One of ten 1982 Spartan/Van Pelt Triples was this one from Fire Station 73 in the San Fernando Valley area of Los Angeles. All ten of these apparatus were equipped with 1500 gpm Hale pumps and 500 gallon water tanks.

Fire Stations 2, 33, 64, 70, and 89 all received one of these 1982 Seagrave 100 foot tillered aerial trucks.

Engine 27's wagon battery and several ladder pipes are in use at this stubborn structure fire.

Units from the LAFD and Los Angeles County Fire Department respond to a brush fire.

The LAFD Supply and Maintenance Section housed this Emergency Air 1, a 1983 Volvo White with an American Bristol air cascade system.

The LAFD was a good customer of the Seagrave Company as it purchased 41 of these 1984 models. Each engine was rated at 1500 gpm and had a 500 gallon water tank.

Three engine companies at work at this greater alarm fire in Los Angeles.

An aggressive attack by LAFD companies during a night fire.

Haz-Mat Squad 4, a converted Ford-LaFrance Triple, at a fire at the county health building in downtown LA.

The 1972 Ford-LaFrance Haz-Mat vehicles were converted again into urban search and rescue rigs.

Haz-Mat Squad 27 was one of three 1972 Ford C-8000 American LaFrance Triples that were converted to hazardous materials rigs in 1978.

Three of these 1985 Seagrave 100 foot aerial trucks were purchased. They were in service at Fire Stations 11,14, and 27.

LAFD Air Operations flew three of these 1984 Bell 206 Jet Rangers which each were capable of carrying five passengers.

The LAFD received seven of these 1986 Seagrave 1500 gpm triples with manual transmissions. Engine 204 was the second engine of a two piece engine company, and part of the task force at Fire Station 4.

Two of these hazardous material vehicles were made by Chevrolet/Suncrest in 1985. They were assigned to Fire Stations 3 and 88.

Crash 90, a 1980 Oshkosh T-6 4x4, was first in service at LAX as Crash 80, but later was assigned to Fire Station 90 at Van Nuys Airport. This unit was known as a Rapid Intervention Vehicle.

Engine 81 rolled out with this 1988 Seagrave 1500 gpm triple with a 500 gallon water tank. Ten of these rigs were on the department's roster.

Engine 228 using its wagon battery to attack a brush fire.

Water Tender 88 was a 1989 Peterbilt six-wheel drive tanker with a 4000 gallon tank.

Water Tender 88 was a former street flusher built by Klein.

This Ford water tanker was stationed at Fire Station 86 in the Studio City area.

Water Tender 86 was used to supply water to the engine companies at brush fires in the hillsides of LA where there are no fire hydrants.

The crew of Engine Company 14 prepares to attack a fully involved structure fire in a commercial district.

Chapter 7

End of the 20th Century, 1990s

The 1990s saw more modern apparatus added to the LAFD fleet. The rescue ambulances were the modular type, the haz-mat squad vehicles were custom built Saulsburys, Hummers were used as brush patrol rigs, Pierce and Seagrave engines were purchased in large numbers, and Emergency One Titan crash trucks were in service at the airport. All of the LAFD apparatus were now purchased with closed cabs.

Ladder pipes working a roof fire at a picture frame store.

This 1990 Seagrave engine was one of seven delivered. It was equipped with a 1500 gpm Waterous pump and carried 500 gallons of water.

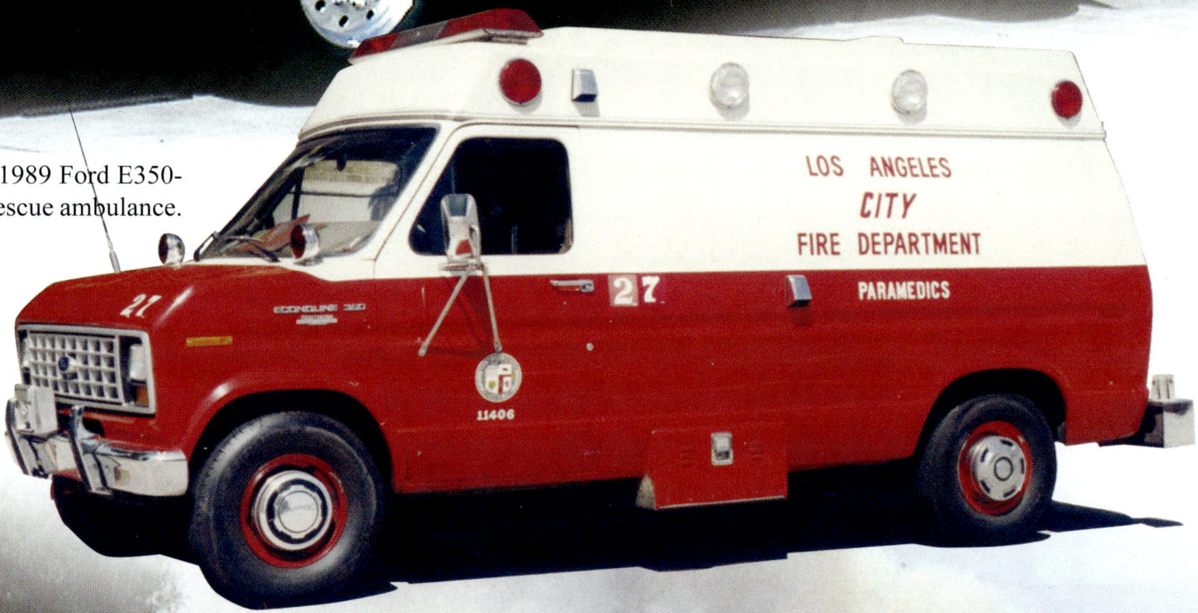

Parked at the curb is a 1989 Ford E350-Southern paramedic rescue ambulance.

The crew of Engine 15 prepares to defend a home against a brush fire.

Oshkosh built this 1990 T-3000 1500 gpm pump with a 3000 gallon water tank assigned as Foam 3 at LAX.

After serving at LAX, this 1990 Oshkosh was painted lime green and reassigned as Foam 114 at Van Nuys Airport Air Operations (Air Ops).

Fire 2 was a 1990 Bell 412 model.

Fire 3, a Bell 412, makes a water drop on a brush fire.

Fire 2 is making a water drop in Griffith Park just below the Observatory.

The Anaheim Fire Department, in nearby Orange County, loaned this 1990 Emergency-One "Quint" with a 75 foot aerial to the LAFD for evaluation. The LAFD rejected the quint concept after a trial period.

Two of these 1990 Seagrave 100 foot aerial ladder trucks served Los Angeles.

When parking gets tight you just have to run the hoses under vehicles.

Haz-Mat Squad 70 operated this 1992 Saulsbury unit in the San Fernando Valley. Three of these squads were delivered to the city. They were assigned to stations 4, 48, and 70.

A quartet of aerials attend to this structure fire in North Hollywood.

This 1994 GMC/SuperVac Light Utility vehicle was an indispensable piece of equipment at night fire scenes. The light tower could be raised for use or lowered for travel.

The LAFD purchased two of these 1995 American/General Hummer 4x4 brush patrol rigs built by Phenix Enterprises. These patrol trucks were powered by Chevrolet 350, V-8 gasoline engines and were deployed to heavy brush areas of the city at Fire Stations 47 and 109.

This rear view of a Hummer Brush Patrol shows the hose arrangement. Both trucks had 100 gpm Hale pumps and 200 gallon water tanks. One of these trucks has been preserved by the LAFD Historical Society.

A total of 42 Seagrave Triples were placed in service between 1995 and 1996. They had 1500 gpm pumps and carried 500 gallons of water.

This 1995 Simon Duplex/LTI 100 foot aerial truck, Truck 35, was in service in the East Hollywood community and was one of 14 purchased by the city in 1995 and 1996.

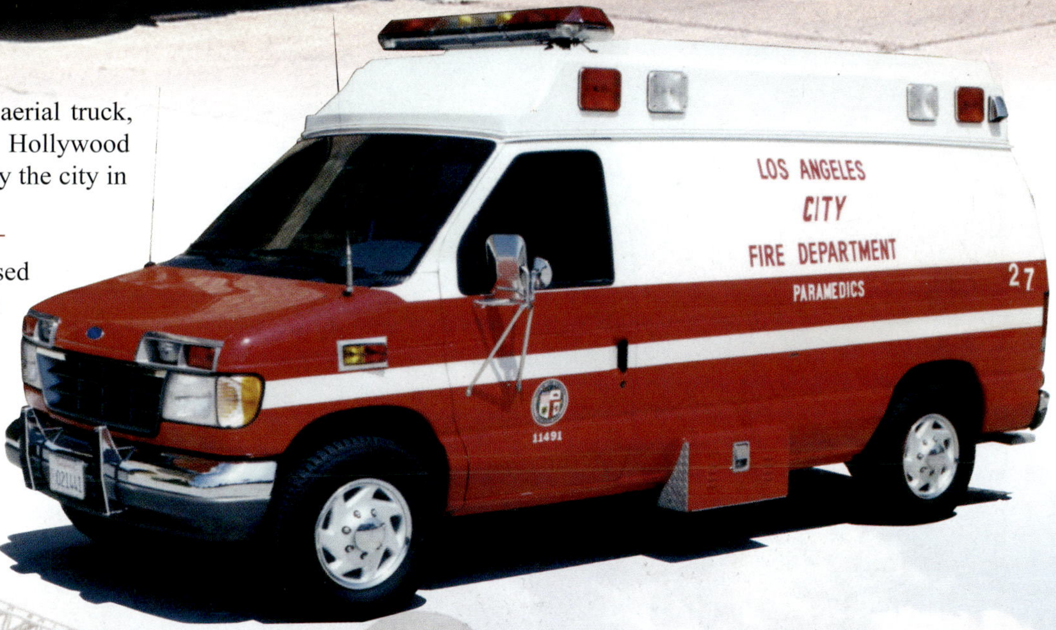

Thirty-three of the ambulances purchased in 1995 were these Ford E350/Southern Navistar 7.3 liter diesel van types.

Truck 35 moved up to a structure fire in the Porter Ranch area. All of the new LAFD aerial ladder trucks had enclosed tillers by the mid 1990s.

In 1995 Heavy Rescue Company 56 received this new Peterbilt 6x4 heavy utility with a 40 ton Century crane.

This rear view shows the 470 hp Heavy Rescue 56 with the stabilizers down and the crane raised.

A 1996 Seagrave Triple at a structure fire in the Valley.

Engine 27, a 1998 Pierce Dash 2000, parading the colors honoring those who died on September 11, 2001.

A raging brush fire approaches Engine 106's crew.

Engine 51, a 1996 Seagrave Triple, was assigned to Fire Station 51 at LAX and is equipped with a foam system.

Dozer Tender 2, a 1999 Ford 450 4x4 with a Pacific body was stationed with the department's bulldozer companies at Fire Station 88.

The LAFD put 27 of these 1998 Pierce Dash 2000 1500 gpm triples in service in 1999. Besides a 500 gallon water tank they also carried 45 gallons of class A foam.

Fire Station 24's 1999 Ford F-350 4x4 brush patrol with a top mounted hose reel and a 200 gallon tank.

This hydrant is getting a test of its capacity with two four-way valves attached.

This 1998 Freightliner carries 2000 gallons of Jet-A fuel to support the helicopter fleet at Van Nuys Airport.

This heli-tender is ready to refuel Fire 2. The wind-sock is attached to the tender.

Foam 1-80 was a 1998 Emergency One Titan 8x8 with a 2000 gpm Hale pump, a 4000 gallon water tank, a 510 gallon foam tank, 700 pounds of Purple K dry chemical and 20 pounds of Halon. The roof turret can put out 1200 gpm. The truck contains a global positioning system that allows it to operate in foggy weather and a heat sensor to locate fire inside aircraft.

LAX was the home for Crash 80, a 1998 Oshkosh T-1500 with a 1500 gpm pump, a 1500 gallon water tank, and a 200 gallon Aqueous Film Forming Foam (AFFF) tank.

This crash truck, which is like Foam 1-80, has an additional 50 foot,1000 gpm Snozzle and was known as Foam 2-80. It was powered by a 1000 hp motor, has a 2000 gpm pump, and carries 4000 gallons of water, 510 gallons of AFFF foam, and 700 pounds of Purple K.

The foam tender in operation at a fire.

This Los Angeles County 1999 KME-Renegade Quint was on loan for evaluation purposes only.

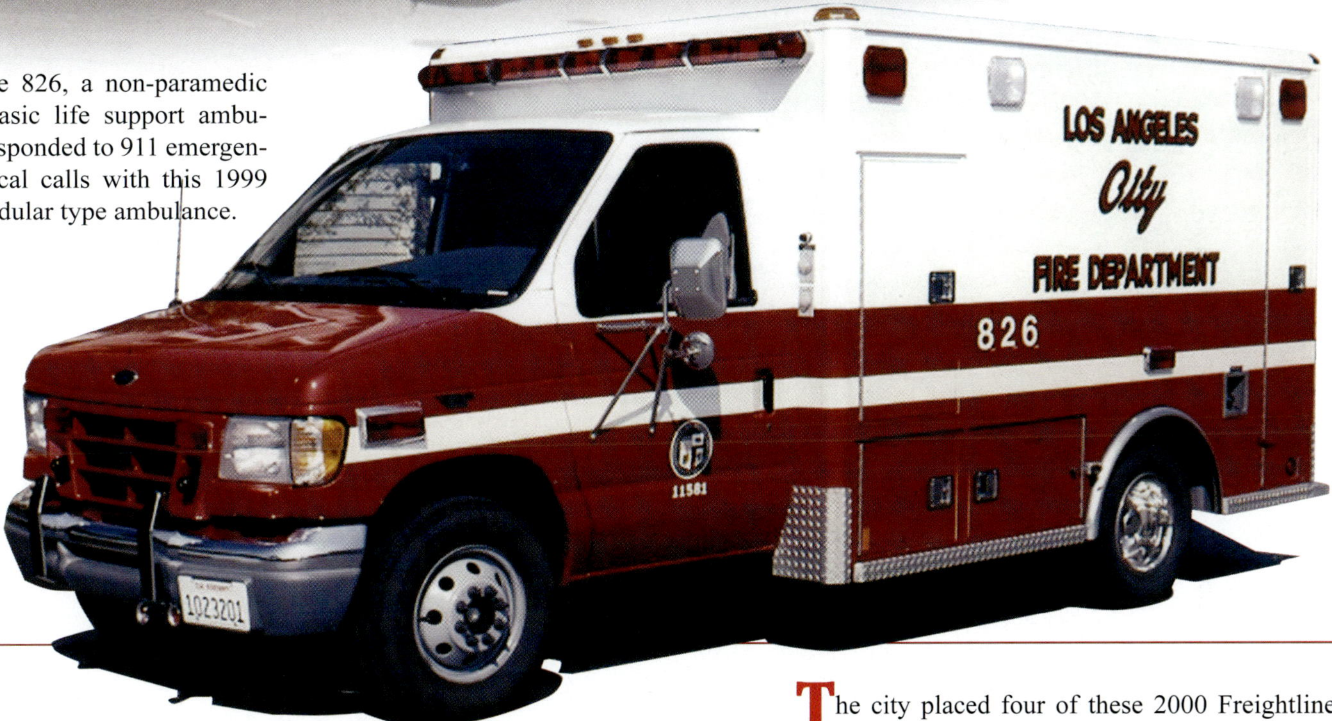

Rescue 826, a non-paramedic or basic life support ambulance, responded to 911 emergency medical calls with this 1999 Ford modular type ambulance.

The city placed four of these 2000 Freightliner-Pierce Foam Tenders in service from San Pedro to the San Fernando Valley.

Several companies of firefighters responding to an alarm on Victory Blvd. in North Hollywood.

This huge 2000 Emergency One foam crash truck, assigned to Fire Station 80 at LAX, is dwarfed by the Quantas airliner.

Foam 80 at a hot drill.

The LAFD placed two of these 2001 Peterbilt 385 tractor transports into service at Fire Station 88. The tractors pull 2001 Murray trailers which carry Caterpillar dozers.

Helicopter Tender 2 carries 3800 gallons of Jet-A fuel for the fleet of helicopters stationed at Fire Station 90 at Van Nuys Airport. Tender 2 was a 2001 Freightliner with a body built by Weld-it of Los Angeles.

A wagon battery in use at a smoky fire in the Produce District.

A major fire in the downtown Fashion District.

Air Stair 80 was a 2001 Ford F-550 air stair truck which was used to quickly evacuate passengers from a plane in the event of an emergency at LAX.

The air stair truck in use at LAX.

Fireboat 1 was one of three 40 foot 2002 boats built by Kvichak Marine Industries in Seattle, Washington. These boats were equipped with large 2400 gpm Hale pumps.

New fire boats 1, 3, and 5 replaced three 36 foot boats built in 1967 by Drake Craft. These new boats were berthed at Fire Stations 111, 49, and 110. Here they are shown with the old boats, Boat 4, and the new large Boat 2 in 2002.

Engine 41 framed by two ladder pipes in action.

A series of wagon batteries in use at a greater alarm fire.

This engine is about to become a fireboat in a flooded intersection.

This structure fire created a street full of apparatus.

A rear view of the Peterbilt Heavy Rescue 56.

The HR in operation lifting a vehicle out of a storm channel.

Peterbilt/Century built this 2003 heavy rescue assigned to Fire Station 56.

The new Fireboat 2, the Warner L. Lawrence, replaced the historic 1925 Ralph J. Scott. The new boat was built by Nichols Brothers Boat Builders in 2002, is 105 feet long, has a pumping capacity of 38,000 gpm, and can travel at a speed of up to 13 knots. The Lawrence is berthed with land companies at Fire Station 112 in San Pedro.

Fireboat 2 with a water display in the main channel at Los Angeles Harbor.

Between the years of 2000 and 2007, the LAFD put 33 of these American LaFrance LTI 100 foot aerial trucks in service.

American LaFrance/Freightliner built four of these 2003 Rehab Air Tenders. They carried spare air bottles, an air compressor, a toilet and shower, food and water, plus a light tower. One of these units was housed at the rescue maintenance section.

This 2003 Ford Rescue Ambulance was assigned to Fire Station 100.

Seventeen of these Seagrave Marauder 1500 gpm triples with 500 gallon water tanks and 30 gallon class A foam tanks were purchased by the city in 2003.

A rear view of a Seagrave Marauder showing the hose beds and enclosed ladder storage.

In 2003, Pierce Arrow delivered 17 of these 1500 gpm triple combination engines with 500 gallon water tanks and 30 gallon class A foam tanks.

A rear view of Engine 7's 2003 Pierce Arrow Triple showing the ladder storage and hose beds.

These fire crews have their work cut out for them at this furniture factory fire in the Valley.

One of three 2005 Zodiac river rafts used for swift water rescues.

The LAFD Swift Water Team responded to water rescues in the rivers and storm channels of Los Angeles with this 2003 Ford F450 four wheel drive. Its body was built by Master Body Works and it has a trailer for the jet skis.

Three of these 2004 Ford/Generals were put into service as Haz-Mat tenders.

The crew of Engine 83 assessing the progress of a fast moving brush fire.

Engine 275 laying a hose line on a mountain road.

Engine 250 using both inlets at a joint agency fire.

Two companies arrive to begin an attack on a fire in the Newhall Pass, north of Los Angeles.

Crew 3 shoulder patch.

The cadet hand crew preparing for action at a fire scene.

This Freightliner Crew Carrier transports members of the LAFD's Crew 3 volunteer cadet hand crew. The truck has been on loan from the L.A. County Fire Department since July, 2007.

These Kawasaki motorcycles were used to transport paramedics during the 405 Freeway closure in 2011.

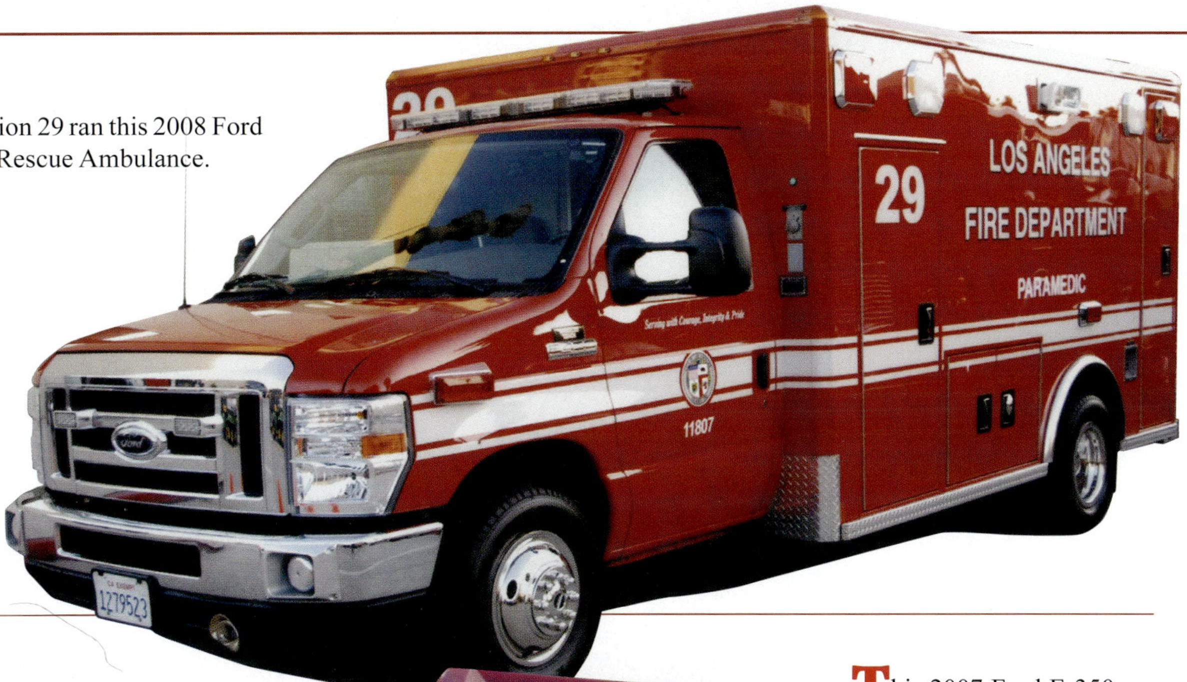

Fire Station 29 ran this 2008 Ford E-450 Rescue Ambulance.

This 2007 Ford E-350 was outfitted as a Battalion Chief's vehicle.

Water Tender 77 is one of two 2008 Peterbilt-KME 500 gpm 2500 gallon mobile water sources that respond to fires in hill-side brush areas.

Engine 5, a 2006 Seagrave Triple, joins the the other companies at the night fire in the Sepulveda Pass.

Engine Co. 63 with their 2006 Pierce Triple at a fire in the Sepulveda Pass, between West Los Angeles and the Valley, in October of 2008.

This greater alarm fire in Chatsworth brought in companies from several nearby stations.

A combined effort of engines and helicopters controls a brush fire next to the 405 Freeway in the Sepulveda Pass in September of 2010.

The Augusta Westland helicopter in action at a hillside fire.

LAFD helicopters making multiple water drops at a night fire in the Sepulveda Pass

Fire 5, a 2008 Augusta Westland 139, is stationed at Air Operations, Fire Station 114 at Van Nuys Airport.

This 2010 KME 1750 gpm triple had a 500 gallon water tank and a 30 gallon class A foam tank. The LAFD purchased 23 of these triples.

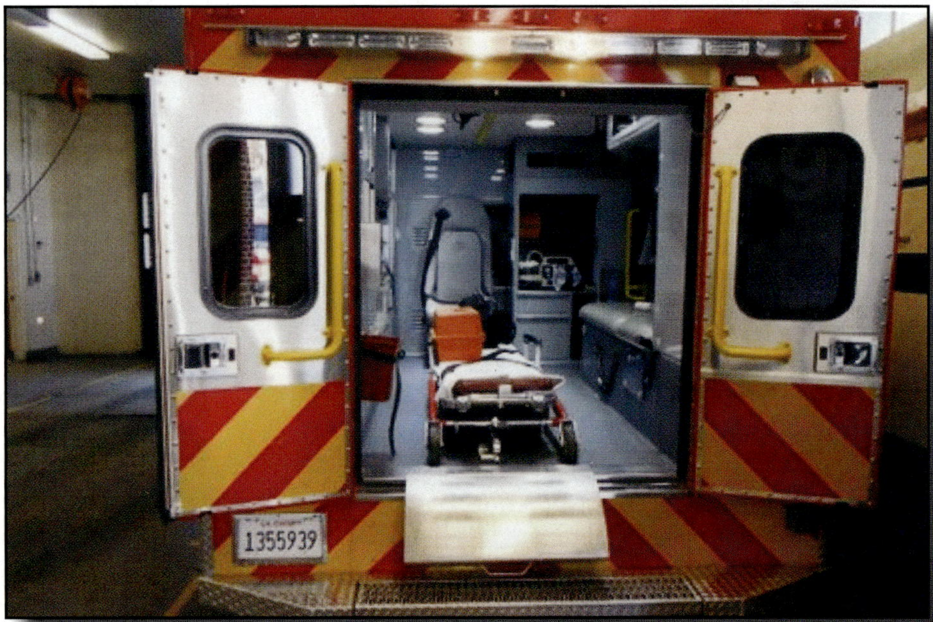

A rear view of a new Dodge ambulance showing the yellow warning stripes on the back of the rig. The patient compartment can be hydraulically lowered for easier loading of the gurney.

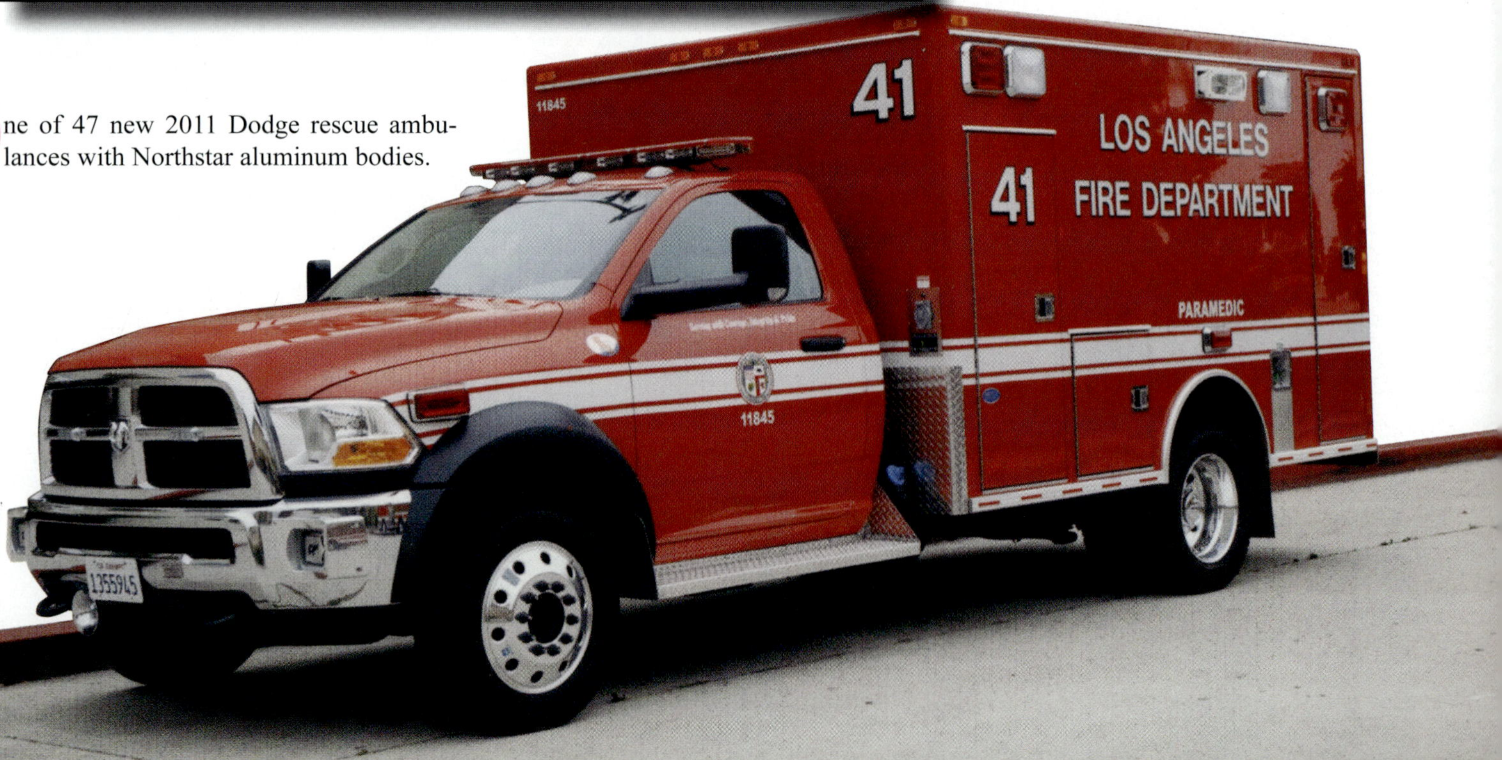

One of 47 new 2011 Dodge rescue ambulances with Northstar aluminum bodies.

1905 Gorter Water Tower, 1:32 scale (3/8" = 1'), by Tom Showers.

Chapter 9

Apparatus Models

Presented here are a selection of miniature scale replica LAFD fire apparatus on display at the Los Angeles Fire Department Museum in Hollywood, California. All the models are entirely scratchbuilt, using wood, styrene plastic, aluminum, and brass.

1915 Moreland Light Wagon 1:32 scale, by John A. Ackerman.

1913 Seagrave tractor attached to a 1910 Anderson City Service Truck, 1:32 scale, by Walt Jaeger.

A modernized 1923 Seagrave Chemical Hose Wagon, 1:32 scale, by John A. Ackerman.

1929 American LaFrance Triple, 1:32 scale, by John A. Ackerman.

1930 Moreland/Standard Auto Body Works Tank Wagon, 1:32 scale, by John A. Ackerman.

Top view of the 1930 Moreland/SABW Tank Wagon.

1941 United-Kenworth Triple, 1:32 scale, by John A. Ackerman.

1937 American LaFrance Duplex pumper, 1:32 scale, by Tom Showers.

1938 American LaFrance Manifold Wagon, 1:32 scale, by Tom Showers.

1938 Seagrave Manifold Wagon, 1:32 scale, by Ian Scott.

Below:
Rear view of the 1938 Seagrave Manifold Wagon.

1956 Seagrave Triple, 1:32 scale, by John A. Ackerman.

1961 Crown Triple, 1:16 scale, by Joe McCloy.

This 1961 Crown model, in 1:16 scale (3/4" = 1'), is 18 inches in length.

1976 Yankee/Walter Crash (LAX), 1:32 scale, by John A. Ackerman.

1971 Seagrave aerial ladder truck, 1:32 scale, by John A. Ackerman.

1968 **Crown 50-foot Snorkel, 1:32 scale, by John A. Ackerman.**

Fireboat 2, the 1925 Ralph J. Scott, 1:24 scale (1/2" = 1'), by Charles Bower. This model is 48 inches in length.

Acknowledgements

To compile this book, every effort was made to use photos that were not printed in any other publication. However, it is inevitable that some photos chosen may have been seen somewhere else or in some other format. As there were hundreds of photos, negatives and slides available to choose from, it was very difficult to select photos that depicted all of the LAFD apparatus while being interesting for the viewer. We regret that we had to leave out so many great photos due to the limited space of this book.

This pictorial history of the apparatus of the Los Angeles City Fire Department was made possible through the efforts of the members of the Los Angeles Fire Department Historical Society and the society's photo archives. We are also grateful to the following photographers and collectors of historic photos: John A. Ackerman, Larry Arnold, George "Smoky" Bass, Dave Cox, Keith D. Cullom (*www.fire-image.com*), Larry Cummings, Phil Glickman, Walt Jaeger, Chuck Madderom (*www.fireapparatusphotography.com*), Frank Manwarren, Mike Meadows (*www.pbase.com*), Don Nash, Joe Ortiz, and the photo archives of the Los Angeles Fire Department.

A special thanks to all members of the LAFD Historical Society, active and retired members of the Los Angeles Fire Department, and friends of the Historical Society Museum who helped with selecting photos, providing information, researching and writing photo captions, proofreading, editing, scanning photos, and support to make this book possible including: John A. Ackerman, Dustin Cornelius, Don Dodd, Jim Finn, and Jennifer Finn.

We have made every effort to credit the individuals, both living and deceased, who have contributed to this project. We apologize for any omissions or errors that may have occurred.

Bibliography

125 Years of Service & Protection: Los Angeles Fire Department, Los Angeles Fire Department Historical Society (LAFDHS) 2011, Faircount Media Group

Crown Firecoach 1951-1985 Photo Archive, Chuck Madderom, Iconografix 2001

Enjine Enjine, Society for the Preservation and Appreciation of Antique Motor Fire Apparatus in America (SPAAMFAA)

Fire Apparatus: A Pictorial History of the LAFD, George Klass, John Ruccione 1975

Firemans Grapevine, Los Angeles Firemen's Relief Association

LAFD Apparatus Records 1872-1998, Larry Schneider, *lafire.com*

LAFD Centennial 1886-1986, Paul Ditzel, Jostens 1986

LAFD Motorized Apparatus 1910-1986, Walt Pittman, Dale Magee, Robert Allen

Land, Sea and Air: A Pictorial History of the LAFD, LAFDHS, M.T. Publishing Company, Inc.™ 2006

Los Angeles City Fire Apparatus 1953-1999, Chuck Madderom, Iconografix 2000

Seagrave, Matthew Lee, Industrial Graphics 1991

Index

A

Ackerman, John A. 3, 199, 200, 201, 203, 204, 206
Air Stair
Ford 174
Air Tender
American LaFrance/Freightliner 180
Air Utility
Mack 108
Volvo 141
Allen, Robert 3, 206
Allis-Chalmers 64
Ambulance
Cadillac 48
Chevrolet 119, 128, 135
Dodge 198
Ford 93, 105, 119, 150, 157, 165, 180, 186, 193
Freightliner-Wheeled Coach 169
GMC 62, 109
International 104
AMC Matador 126
American Bristol 141
American LaFrance 5, 12, 25, 26, 27, 32, 65, 71, 121, 135, 144, 180, 200, 202
American LaFrance/LTI (Ladder Towers Inc.) 169
Anderson City Service 12
Arnold, Larry 3, 206
Arson Investigation
Ford 72
Augusta-Westland 169, 196, 197

B

Bass, George "Smoky" 206
Bell 132
Boat Tender
Mack/Yankee 43
Bower, Charles 205
Brush Tanker
Moreland 21
Buick 28, 38, 61

C

Cadillac 48
Cat 100
Caterpiller 98, 102, 110
Chemical Engine 10, 11
Chemical Hose Wagon 7, 11
Fire Extinguisher Co. 10
Racine 11
Seagrave 13, 20, 200
Stutz 15
Chevrolet 78, 85, 119, 145, 156
Christie 12
Combination Engine
American LaFrance 12
Seagrave 16
Command Vehicle 187
AMC Matador 126
Buick 28, 38, 61
Chevrolet 190
Ford 193
Freightliner 186
Plymouth 129
Pontiac 128
Communications
Chevrolet 188
Ford 56, 73
Cornelius, Dustin 206
Cox, Dave 206
Crash Truck
Crown 127
Emergency One 149
FWD/Yankee 54, 55, 74, 87
International 33
Mack 30
Oshkosh 134, 145, 163
Oshkosh/Yankee 136
Walter/Yankee 88, 98, 136, 204
Crown 48, 53, 56, 57, 59, 63, 64, 69, 76, 77, 80, 81, 85, 86, 89, 91, 95, 97, 99, 100, 101, 103, 104, 106, 109, 110, 114, 115, 116, 117, 118, 120, 121, 123, 124, 125, 126, 127, 203, 205, 206
Crown Firecoach 5, 63
Cullom, Keith 3, 206
Cummings, Larry 206

D

Dempsey, Leo 14
Ditzel, Paul 3, 206
Dodd, Don 206
Dodge 83, 94
Dozer Tender
Ford 160
Duplex Pumpers
American LaFrance 202
American LaFrance Metropolitan 13

E

Emergency One 149, 163, 170

F

Finn, Jim 206, Jennifer 206
Fire Boat 7
Boat 1 63, 94, 175
Boat 2 7, 18, 19, 96, 175, 179, 205
Boat 3 63, 96, 175
Boat 4 63, 72, 96, 106, 111, 175
Boat 5 63, 96, 175
Fire Extinguisher Co. 10
Foam Truck 130, 164, 166, 171
Crown 57, 130
Emergency One 163, 170
Ford 75
Freightliner-Pierce 165
GMC 13, 135
Moreland 17
Oshkosh 151
Ford 56, 60, 66, 72, 73, 74, 75, 91, 93, 119, 121, 133, 143, 144, 148, 150, 154, 157, 160, 161, 165, 166, 174, 180, 183, 186, 193
Ford-LaFrance 143
Freightliner 162, 165, 169, 172, 192
Fuel Truck 102
GMC 34, 83
FWD/Yankee 54, 55

G

Glickman, Phil 206
GMC 34, 44, 46, 47, 65, 79, 80, 83, 90, 91, 97, 101, 109
Gorter 69, Henry 7
Griffin, Tim 14

H

Haines Roadster 7
Hall-Scott 20, 65, 101
Hayes, Daniel 7
Haz-Mat 135, 143, 144, 154, 183, 187
Chevrolet/Suncreast 145
Ford/Generals 183
Pierce Arrow 186
Saulsbury 154
Heavy Rescue
Peterbilt 158, 178
Heavy Utility 61
Crown 100, 101, 125
Kenworth 37, 98
Mack 42, 110, 125
Mack/Holms 123
Peterbilt 172
Helecopter Tender
Ford 60
Freightliner 162, 172
Mack 133
Helicopter 122
Augusta Westland 196, 197
Bell 78, 79, 87, 122, 132, 144, 152
Chinook 138
Erickson Air-Crane 167
Holmes 37
Hose Wagon 62
Crown 56, 130
Crown Firecoach 39
Mack 20, 39
Seagrave 22
Hotchkin, Frank 136
Howell, Mark 14

I

International 5, 33, 34, 53, 55, 64

J

Jaeger, Walt 200, 206
John Deere 169, 190

K

Kawasaki 192
Kenworth 5, 33, 37, 40, 98
Kimball, Ward 21
Klass, George 206
Klein 147
KME (Kovatch Mobile Equipment) 5, 169, 193, 198
Kuhrts, Jacob 9
Kvichak Marine Industries 175

L

Ladder Truck 18
Aerial 154, 157, 189
American LaFrance 13, 32, 180
Crown/Maxim 104
Emergency One 153
Hayes 7
Pirsch 40, 62, 71
Seagrave 17, 32, 39, 45, 50, 51, 67, 68, 73, 77, 82, 90, 92, 109, 113, 120, 125, 129, 132, 140, 144, 153, 204
Simon Duplex/LTI 157
Thibault 113, 117
City Service
American LaFrance 24, 29
Anderson 12
Stutz 16
LaFrance, American 24
Lee, Matthew 206
Light Utility
GMC 46
Moreland 199

M

Mack 5, 20, 30, 38, 39, 41, 42, 108, 110, 133
Madderom, Chuck 3, 206
Magee, Dale 3, 206
Manifold-Duplex System 13, 63

Index

Manifold Wagons
American LaFrance 13, 25, 27, 202
Crown 85, 86
Mack 13, 93
Seagrave 13, 26, 27, 202
Manwarren, Frank 206
McCloy, Joe 203
Meadows, Mike 2, 3, 206
Moreland 5, 11, 17, 20, 21, 199, 201
Moreland Rescue Squad 13

N

Nash, Don 206
Nott 10

O

Ortiz, Joe 206
Oshkosh 52, 145, 151, 163
Oshkosh/Yankee 136

P

Patrol Truck
American/General Hummer 156
Dodge 83
Ford 93, 133, 161, 166
GMC/Yankee 90
International 34
Personnel Carrier
Freightliner 192
GMC 97
Moreland 20
Peterbilt 147, 158, 178, 193
Pierce 5, 149, 159, 195
Pierce Arrow 182, 186, 188, 190
Pierce Dash 159, 160
Pirsch 5, 31, 40, 62, 71
Pittman, Walt 3, 206
Pumper
American LaFrance 21, 26, 71
Crown 53, 57, 103

Q

Quad
United-Kenworth 29
Quint 153
KME-Renegade 165

R

Racine 11
Rapid Intervention Vehicle
Oshkosh 145
Walter/Yankee 131
Ruccione, John 206

S

Salvage Wagon
Chevrolet/Yankee 85
Dodge/Yankee 94
Ford/Yankee 91
GMC 34, 65
GMC/Yankee 91
International 64
Mack 20, 108
Stutz 15
Schneider, Larry 206
Scott, Ian 202
Seagrave 5, 12, 14, 16, 17, 20, 22, 26, 27, 32, 33, 37, 39, 45, 49, 50, 51, 59, 67, 68, 73, 77, 82, 84, 90, 92, 107, 109, 113, 117, 120, 123, 125, 129, 132, 137, 139, 140, 141, 144, 145, 146, 149, 150, 153, 155, 156, 158, 181, 187, 191, 194, 200, 202, 203, 204, 206
Search and Rescue Trailer 83
Service Utility
Chevrolet 78
GMC 101
Showers, Tom 199, 202
Snorkel 99, 110
Crown 99, 106, 110, 114, 116, 117, 205
Spartan/Van Pelt 139
Squirts
Crown 115, 120
Stang 80
Steamer 7
American Metropolitan 2, 8, 12
Amoskeag 5, 9
Nott 10
Stutz 15, 16
Suncrest 145
Swift Water Rescue
Ford 183

T

Tank Wagon
GMC 44, 45, 80, 128
GMC/Yankee 79
Kenworth 40
Kenworth/Yankee 47
Mack 41
Moreland/SABW 201
Oshkosh 52
Telesquirt
Crown 115, 118
Thibault 113, 117
Tractor
American LaFrance 13
Caterpillar 97
Christie 12
Crown 64
Seagrave 12, 200
Triple Combination Pumper
Crown 116
Triple Engine
American LaFrance 24, 65, 121, 143, 144, 200
Crown 46, 48, 59, 69, 77, 80, 81, 86, 89, 95, 97, 103, 109, 120, 121, 124, 126, 139, 203
Crown Firecoach 39
Kenworth 33
KME 198
Mack 38, 41, 59
Pierce 195
Pierce Arrow 182, 190
Pierce Dash 159
Pirsch 31
Pirsch Dash 160, 168
Seagrave 2, 13, 14, 15, 32, 33, 37, 49, 59, 84, 107, 137, 139, 141, 145, 146, 150, 156, 158, 160, 194
Seagrave Marauder 181, 187, 191
Spartan/Van Pelt 139
United-Kenworth 28, 31, 201
Van Pelt 111
Ward LaFrance 129

U

United-Kenworth 28, 29, 31
Urban Search and Rescue 189
Pierce Arrow 188
Pierce Dash 168
Utility
Allis-Chambers Bulldozer 64
Catepiller Bulldozer 98, 102, 110, 172
Ford Commissary 66, 74
Ford Tow Truck 75
GMC 47
GMC/SuperVac (light) 155
International 53, 102
International Dump Truck 137

V

Van Pelt 111

W

Wagon Battery 172
Walter 88, 204
Ward-LaFrance 129, 131
Water Tender
Ford 148
Peterbilt 147
Peterbilt-KME 193
Water Tower 7
American LaFrance 28, 29
Gorter 7, 13, 69, 199
White Rescue Squad 13

Y

Yankee 74, 79, 85, 87, 88, 90, 91, 94, 204
Yankee/Walter 204

LIBROS
RITMO ROCK
ARCADE
Ritmo LATINO
cingular WIRELESS
Ritmo LATINO
DolEx
Mercedes
VOCATIONAL
SCHOOL
ELVIRA'S
WEDDING CHAPEL
CASAMIENTOS
INMEDIATOS
(213) 620-8886
Mercedes VOCATIONAL SCHOOL
BRIDAL & TUXEDOS
INCOME TAX
McDonald's
UNION
STORE FOR MEN
CREDITO FACIL
LOS ANGELES
FIRE DEPARTMENT
827